STUDENT WOR
VOLUME 3

MW00565485

P H Y S I C S

FOR SCIENTISTS AND ENGINEERS SECOND EDITION

A STRATEGIC APPROACH

Randall D. Knight
California Polytechnic State University
San Luis Obispo

PEARSON
Addison
Wesley

San Francisco   Boston   New York
Capetown   Hong Kong   London   Madrid   Mexico City
Montreal   Munich   Paris   Singapore   Sydney   Tokyo   Toronto

Publisher: Adam Black, Ph.D.
Development Manager: Michael Gillespie
Development Editor: Alice Houston, Ph.D.
Project Editor: Martha Steele
Assistant Editor: Grace Joo
Media Producer: Deb Greco
Sr. Administrative Assistant: Cathy Glenn
Director of Marketing: Christy Lawrence
Executive Marketing Manager: Scott Dustan
Sr. Market Development Manager: Josh Frost
Market Development Associate: Jessica Lyons
Managing Editor: Corinne Benson
Production Supervisor: Nancy Tabor
Production Service: WestWords PMG
Illustrations: Precision Graphics
Text Design: Seventeenth Street Studios and WestWords PMG
Cover Design: Yvo Riezebos Design and Seventeenth Street Studios
Manufacturing Manager: Pam Augspurger
Text and Cover Printer: Edwards Brothers
Cover Image: Composite illustration by Yvo Riezebos Design; photo of spring by Bill Frymire/Masterfile

ISBN-13: 978-0-321-51628-2
ISBN-10: 0-321-51628-1

Table of Contents

Preface

Learning physics, just as learning any skill, requires regular practice of the basic techniques. That is what this *Student Workbook* is all about. The workbook consists of exercises that give you an opportunity to practice the ideas and techniques presented in the textbook and in class. These exercises are intended to be done on a daily basis, right after the topics have been discussed in class and are still fresh in your mind.

You will find that the exercises are nearly all *qualitative* rather than *quantitative*. They ask you to draw pictures, interpret graphs, use ratios, write short explanations, or provide other answers that do not involve significant calculations. The purpose of these exercises is to help you develop the basic thinking tools you'll later need for quantitative problem solving. Successful completion of the workbook exercises will prepare you to tackle the more quantitative end-of-chapter homework problems in the textbook. It is highly recommended that you do the workbook exercises *before* starting the end-of-chapter problems.

You will find that the exercises in this workbook are keyed to specific sections of the textbook in order to let you practice the new ideas introduced in that section. You should keep the text beside you as you work and refer to it often. You will usually find Tactics Boxes, figures, or examples in the textbook that are directly relevant to the exercises. When asked to draw figures or diagrams, you should attempt to draw them so that they look much like the figures and diagrams in the textbook.

Because the exercises go with specific sections of the text, you should answer them on the basis of information presented in *just* that section (and prior sections). You may have learned new ideas in Section 7 of a chapter, but you should not use those ideas when answering questions from Section 4. There will be ample opportunity in the Section 7 exercises to use that information there.

You will need a few "tools" to complete the exercises. Many of the exercises will ask you to *color code* your answers by drawing some items in black, others in red, and yet others in blue. You need to purchase a few colored pencils to do this. The author highly recommends that you work in pencil, rather than ink, so that you can easily erase. Few people produce work so free from errors that they can work in ink! In addition, you'll find that a small, easily carried six-inch ruler will come in handy for drawings and graphs.

As you work your way through the textbook and this workbook, you will find that physics is a way of *thinking* about how the world works and why things happen as they do. We will be interested primarily in finding relationships and seeking explanations, only secondarily in computing numerical answers. In many ways, the thinking tools developed in this workbook are what the course is all about. If you take the time to do these exercises regularly and to review the answers, in whatever form your instructor provides them, you will be well on your way to success in physics.

To the instructor: The exercises in this workbook can be used in many ways. You can have students work on some exercises in class as part of an active-learning strategy. Or you can do the same in recitation sections or laboratories. This approach allows you to discuss the answers immediately, to answer student questions, and to improvise follow-up exercises when needed. Having the students work in small groups (2 to 4 students) is highly recommended.

Alternatively, the exercises can be assigned as homework. The pages are perforated for easy tear-out, and the page breaks are in logical places so that you can assign the sections of a chapter that you would likely cover in one day of class. Exercises should be assigned immediately after presenting the relevant information in class and should be due at the beginning of the next class. Collecting them at the beginning of class, then going over two or three that are likely to cause difficulty, is an effective means of quickly reviewing major concepts from the previous class and launching a new discussion.

If the exercisees are used as homework, it is *essential* for students to receive *prompt* feedback. Ideally this would occur by having the exercises graded, with written comments, and returned at the next class meeting. Posting the answers on a course website also works. Lack of prompt feedback can negate much of the value of these exercises. Placing similar qualitative/ graphical questions on quizzes and exams, and telling students at the beginning of the term that you will do so, encourages students to take the exercises seriously and to check the answers.

The author has been successful with assigning *all* exercises in the workbook as homework, collecting and grading them every day through Chapter 4, then collecting and grading them on about one-third of subsequent days on a random basis. Student feedback from end-of-term questionnaires reveals three prevalent attitudes toward the workbook exercises:

i.   They think it is an unreasonable amount of work.

ii. They agree that the assignments force them to keep up and not get behind.

iii. They recognize, by the end of the term, that the workbook is a valuable learning tool.

However you choose to use these exercises, they will significantly strengthen your students' conceptual understanding of physics.

Following the workbook exercises are optional Dynamics Worksheets, Momentum Worksheets, and Energy Worksheets for use with end-of-chapter problems in Parts I and II of the textbook. Their use is recommended to help students acquire good problem-solving habits early in the course. End-of-chapter problems marked with the 🖉 icon are intended to be done on worksheets.

Answers to all workbook exercises are provided as pdf files on the *Media Manager*. The author gratefully acknowledges the careful work of answer writers Professor James H. Andrews of Youngstown State University and Rebecca Sobinovsky.

Acknowledgments: Many thanks to Martha Steele at Addison-Wesley and to Jared Sterzer at WestWords PMG for handling the logistics and production of the *Student Workbook*.

20 Traveling Waves

20.1 The Wave Model

1. a. In your own words, define what a *transverse wave* is.

   b. Give an example of a wave that, from your own experience, you know is a transverse wave. What observations or evidence tells you this is a transverse wave?

2. a. In your own words, define what a *longitudinal wave* is.

   b. Give an example of a wave that, from your own experience, you know is a longitudinal wave. What observations or evidence tells you this is a longitudinal wave?

3. Three wave pulses travel along the same string. Rank in order, from largest to smallest, their wave speeds v_1, v_2, and v_3.

   Order:

   Explanation:

20.2 One-Dimensional Waves

4. A wave pulse travels along a string at a speed of 200 cm/s. What will be the speed if:
   Note: Each part below is independent and refers to changes made to the original string.
   a. The string's tension is doubled?

   b. The string's mass is quadrupled (but its length is unchanged)?

   c. The string's length is quadrupled (but its mass is unchanged)?

   d. The string's mass and length are both quadrupled?

5. This is a history graph showing the displacement as a function of time at one point on a string. Did the displacement at this point reach its maximum of 2 mm *before* or *after* the interval of time when the displacement was a constant 1 mm? Explain how you interpreted the graph to answer this question.

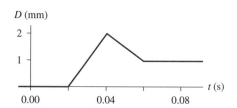

6. Each figure below shows a snapshot graph at time $t = 0$ s of a wave pulse on a string. The pulse on the left is traveling to the right at 100 cm/s; the pulse on the right is traveling to the left at 100 cm/s. Draw snapshot graphs of the wave pulse at the times shown next to the axes.

a.

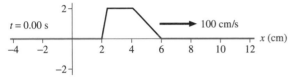

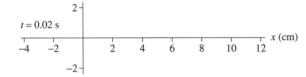

b.

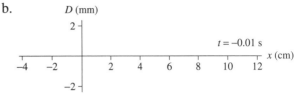

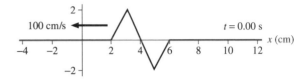

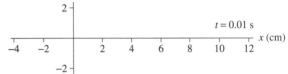

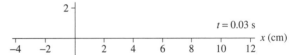

7. This snapshot graph is taken from Exercise 6a. On the axes below, draw the *history* graphs $D(x = 2$ cm$, t)$ and $D(x = 6$ cm$, t)$, showing the displacement at $x = 2$ cm and $x = 6$ cm as functions of time. Refer to your graphs in Exercise 6a to see what is happening at different instants of time.

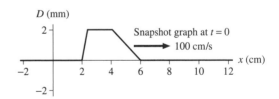

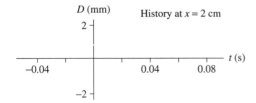

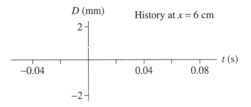

8. This snapshot graph is from Exercise 6b.

   a. Draw the history graph $D(x = 0$ cm, $t)$ for this wave at the point $x = 0$ cm.

   b. Draw the *velocity*-versus-time graph for the piece of the string at $x = 0$ cm. Imagine painting a dot on the string at $x = 0$ cm. What is the velocity of this dot as a function of time as the wave passes by?

   c. As a wave passes through a medium, is the speed of a particle in the medium the same as or different from the speed of the wave through the medium? Explain.

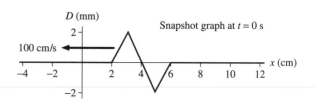

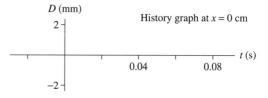

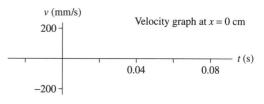

9. Below are four snapshot graphs of wave pulses on a string. For each, draw the history graph at the specified point on the x-axis. No time scale is provided on the t-axis, so you must determine an appropriate time scale and label the t-axis appropriately.

a.
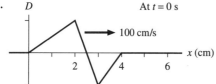

b.
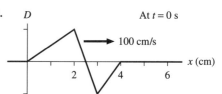

c.
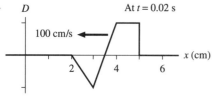

d.
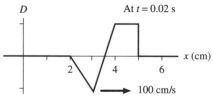

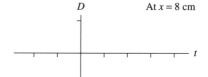

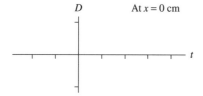

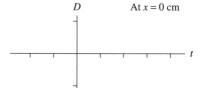

10. A history graph $D(x = 0 \text{ cm}, t)$ is shown for the $x = 0$ cm point on a string. The pulse is moving to the right at 100 cm/s.

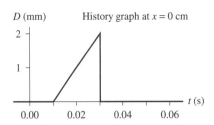

a. Does the $x = 0$ cm point on the string rise quickly and then fall slowly, or rise slowly and then fall quickly? Explain.

b. At what time does the leading edge of the wave pulse arrive at $x = 0$ cm?

c. At $t = 0$ s, how far is the leading edge of the wave pulse from $x = 0$ cm? Explain.

d. At $t = 0$ s, is the leading edge to the right or to the left of $x = 0$ cm?

e. At what time does the trailing edge of the wave pulse leave $x = 0$ cm?

f. At $t = 0$ s, how far is the trailing edge of the pulse from $x = 0$ cm? Explain.

g. By referring to the answers you've just given, draw a snapshot graph $D(x, t = 0 \text{ s})$ showing the wave pulse on the string at $t = 0$ s.

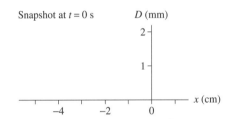

11. These are a history graph *and* a snapshot graph for a wave pulse on a string. They describe the same wave from two perspectives.

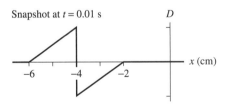 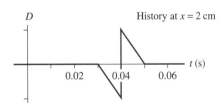

a. In which direction is the wave traveling? Explain.

b. What is the speed of this wave?

12. Below are two history graphs for wave pulses on a string. The speed and direction of each pulse are indicated. For each, draw the snapshot graph at the specified instant of time. No distance scale is provided, so you must determine an appropriate scale and label the *x*-axis appropriately.

a.

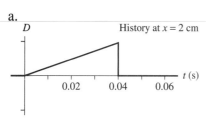

100 cm/s to the left

b.

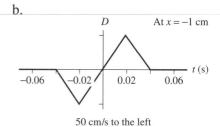

50 cm/s to the left

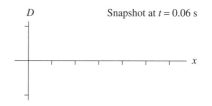

13. A horizontal Slinky is at rest on a table. A wave pulse is sent along the Slinky, causing the top of link 5 to move *horizontally* with the displacement shown in the graph.

a. Is this a transverse or a longitudinal wave? Explain.

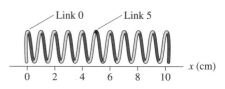

b. What is the position of link 5 at $t = 0.1$ s? _____

What is the position of link 5 at $t = 0.2$ s? _____

What is the position of link 5 at $t = 0.3$ s? _____

Note: *Position*, not displacement.

c. Draw a velocity-versus-time graph of link 5. Add an appropriate scale to the vertical axis.

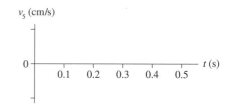

d. Can you determine, from the information given, whether the wave pulse is traveling to the right or to the left? If so, give the direction and explain how you found it. If not, why not?

e. Can you determine, from the information given, the speed of the wave? If so, give the speed and explain how you found it. If not, why not?

14. We can use a series of dots to represent the positions of the links in a Slinky. The top set of dots shows a Slinky in equilibrium with a 1 cm spacing between the links. A wave pulse is sent down the Slinky, traveling to the right at 10 cm/s. The second set of dots shows the Slinky at $t = 0$ s. The links are numbered, and you can measure the displacement Δx of each link.

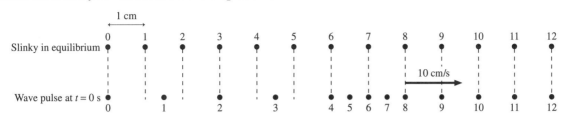

a. Draw a snapshot graph showing the displacement of each link at $t = 0$ s. There are 13 links, so your graph should have 13 dots. Connect your dots with lines to make a continuous graph.

b. Is your graph a "picture" of the wave or a "representation" of the wave? Explain.

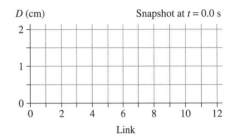

c. Draw graphs of displacement versus the link number at $t = 0.1$ s and $t = 0.2$ s.

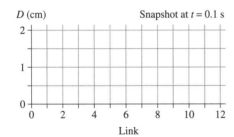

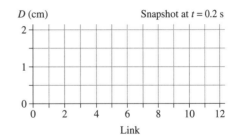

d. Now draw dot pictures of the links at $t = 0.1$ s and $t = 0.2$ s. The equilibrium positions and the $t = 0$ s picture are shown for reference.

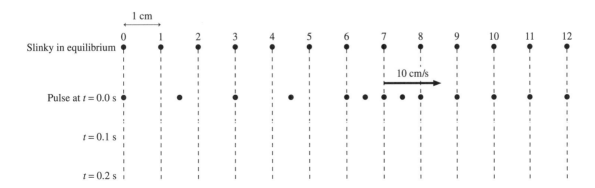

15. The graph shows displacement versus the link number for a wave pulse on a Slinky. Draw a dot picture showing the Slinky at this instant of time. A picture of the Slinky in equilibrium, with 1 cm spacings, is given for reference.

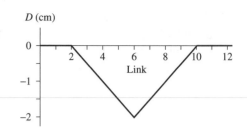

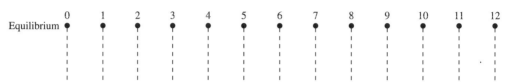

16. The graph shows displacement versus the link number for a wave pulse on a Slinky. Draw a dot picture showing the Slinky at this instant of time. A picture of the Slinky in equilibrium, with 1 cm spacings, is given for reference.

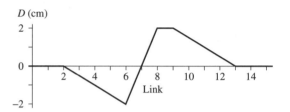

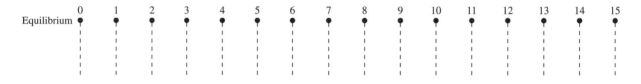

20.3 Sinusoidal Waves

17. The figure shows a sinusoidal traveling wave. Draw a graph of the wave if:

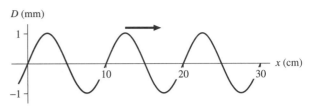

   a. Its amplitude is halved and its wavelength is doubled.

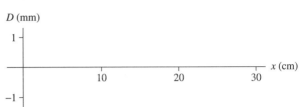

   b. Its speed is doubled and its frequency is quadrupled.

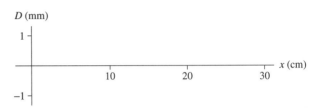

18. The wave shown at time $t = 0$ s is traveling to the right at a speed of 25 cm/s.

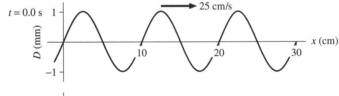

   a. Draw snapshot graphs of this wave at times $t = 0.1$ s, $t = 0.2$ s, $t = 0.3$ s, and $t = 0.4$ s.

   b. What is the wavelength of the wave?

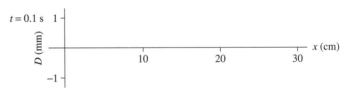

   c. Based on your graphs, what is the period of the wave?

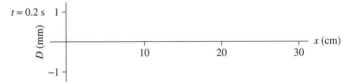

   d. What is the frequency of the wave?

   e. What is the value of the product λf?

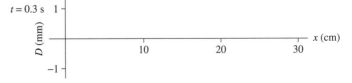

   f. How does this value of λf compare to the speed of the wave?

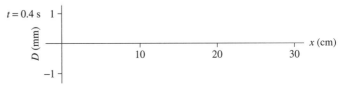

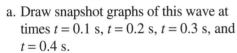

19. Three waves traveling to the right are shown below. The first two are shown at $t = 0$, the third at $t = T/2$. What are the phase constants ϕ_0 of these three waves?

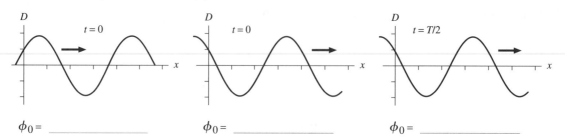

$\phi_0 = $ _____ $\phi_0 = $ _____ $\phi_0 = $ _____

Note: Knowing the displacement $D(0,0)$ is a *necessary* piece of information for finding ϕ_0 but is not by itself enough. The first two waves above have the same value for $D(0,0)$ but they do *not* have the same ϕ_0. You must also consider the overall shape of the wave.

20. A sinusoidal wave with wavelength 2 m is traveling along the x-axis. At $t = 0$ s the wave's phase at $x = 2$ m is $\pi/2$ rad.

   a. Draw a snapshot graph of the wave at $t = 0$ s.

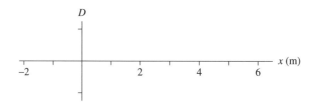

   b. At $t = 0$ s, what is the phase at $x = 0$ m? _____

   c. At $t = 0$ s, what is the phase at $x = 1$ m? _____

   d. At $t = 0$ s, what is the phase at $x = 3$ m? _____

   Note: No calculations are needed. Think about what the phase *means* and utilize your graph.

21. Consider the wave shown. Redraw this wave if:

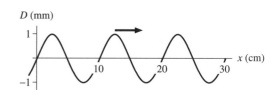

   a. Its wave number is doubled.   b. Its wave number is halved.

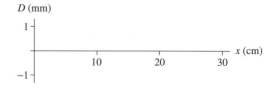

20.4 Waves in Two and Three Dimensions

22. A wave-front diagram is shown for a sinusoidal plane wave at time $t = 0$ s. The diagram shows only the xy-plane, but the wave extends above and below the plane of the paper.

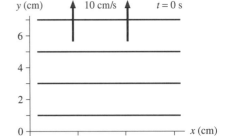

 a. What is the wavelength of this wave? _____

 b. At $t = 0$ s, for which values of y is the wave a crest?

 c. At $t = 0$ s, for which values of y is the wave a trough?

 d. Can you tell if this is a transverse or a longitudinal wave? If so, which is it and how did you determine it? If not, why not?

 e. How does the displacement at the point $(x, y, z) = (6, 5, 0)$ compare to the displacement at the point $(2, 5, 0)$? Is it more, less, the same, or is there no way to tell? Explain.

 f. On the left axes below, draw a snapshot graph $D(y, t = 0$ s$)$ along the y-axis at $t = 0$ s.

 g. On the right axes below, draw a wave-front diagram at time $t = 0.3$ s.

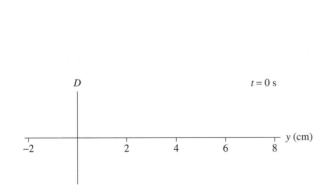

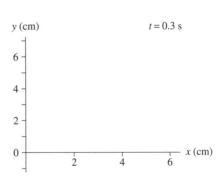

23. These are the wave fronts of a circular wave. What is the phase difference between:

 a. Points A and B? _____

 b. Points C and D? _____

 c. Points E and F? _____

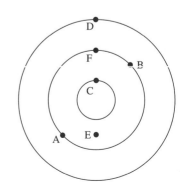

20.5 Sound and Light

24. Rank in order, from largest to smallest, the wavelengths λ_1 to λ_3 for sound waves having frequencies $f_1 = 100$ Hz, $f_2 = 1000$ Hz, and $f_3 = 10,000$ Hz.

Order:

Explanation:

25. A light wave travels from vacuum, through a transparent material, and back to vacuum. What is the index of refraction of this material? Explain.

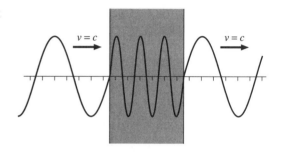

26. A light wave travels from vacuum, through a transparent material whose index of refraction is $n = 2.0$, and back to vacuum. Finish drawing the snapshot graph of the light wave at this instant.

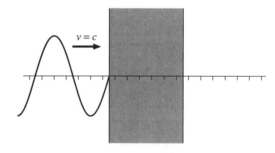

20.6 Power, Intensity, and Decibels

27. A laser beam has intensity I_0.

    a. What is the intensity, in terms of I_0, if a lens focuses the laser beam to $\frac{1}{10}$ its initial diameter?

    b. What is the intensity, in terms of I_0, if a lens defocuses the laser beam to 10 times its initial diameter?

28. Sound wave A delivers 2 J of energy in 2 s. Sound wave B delivers 10 J of energy in 5 s. Sound wave C delivers 2 mJ of energy in 1 ms. Rank in order, from largest to smallest, the sound powers P_A, P_B, and P_C of these three sound waves.

    Order:

    Explanation:

29. A giant chorus of 1000 male vocalists is singing the same note. Suddenly 999 vocalists stop, leaving one soloist. By how many decibels does the sound intensity level decrease? Explain.

20.7 The Doppler Effect

30. You are standing at $x = 0$ m, listening to a sound that is emitted at frequency f_0. The graph shows the frequency you hear during a four-second interval. Which of the following describes the sound source?

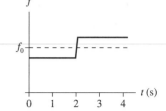

    i.   It moves from left to right and passes you at $t = 2$ s.
    ii. It moves from right to left and passes you at $t = 2$ s.
    iii. It moves toward you but doesn't reach you. It then reverses direction at $t = 2$ s.
    iv. It moves away from you until $t = 2$ s. It then reverses direction and moves toward you but doesn't reach you.

    Explain your choice.

31. You are standing at $x = 0$ m, listening to a sound that is emitted at frequency f_0. At $t = 0$ s, the sound source is at $x = 20$ m and moving toward you at a steady 10 m/s. Draw a graph showing the frequency you hear from $t = 0$ s to $t = 4$ s. Only the shape of the graph is important, not the numerical values of f.

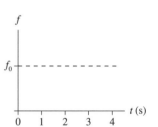

32. You are standing at $x = 0$ m, listening to seven identical sound sources. At $t = 0$ s, all seven are at $x = 343$ m and moving as shown below. The sound from all seven will reach your ear at $t = 1$ s.

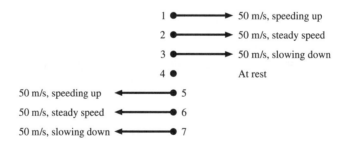

    Rank in order, from highest to lowest, the seven frequencies f_1 to f_7 that you hear at $t = 1$ s.

    Order:

    Explanation:

21 Superposition

21.1 The Principle of Superposition

1. Two pulses on a string, both traveling at 10 m/s, are approaching each other. Draw snapshot graphs of the string at the three times indicated.

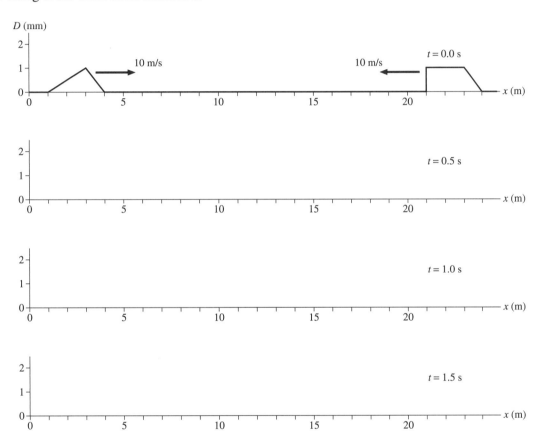

2. Two pulses on a string, both traveling at 10 m/s, are approaching each other. Draw a snapshot graph of the string at $t = 1$ s.

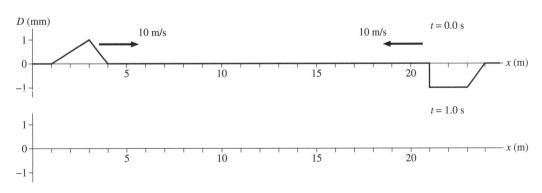

21.2 Standing Waves

21.3 Transverse Standing Waves

3. Two waves are traveling in opposite directions along a string. Each has a speed of 1 cm/s and an amplitude of 1 cm. The first set of graphs below shows each wave at $t = 0$ s.

   a. On the axes at the right, draw the superposition of these two waves at $t = 0$ s.

   b. On the axes at the left, draw each of the two displacements every 2 s until $t = 8$ s. The waves extend beyond the graph edges, so new pieces of the wave will move in.

   c. On the axes at the right, draw the superposition of the two waves at the same instant.

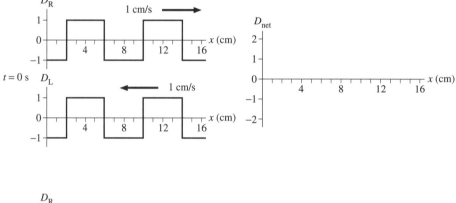

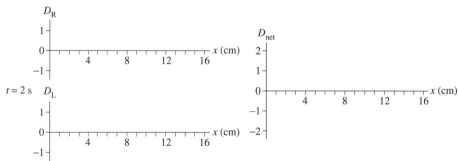

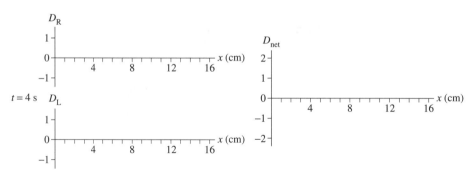

(Continues next page)

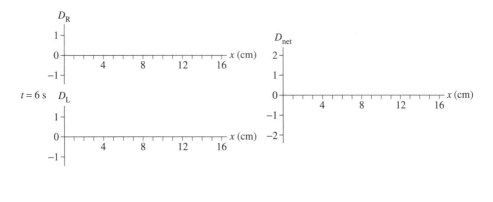

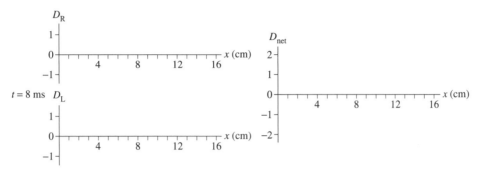

4. The figure shows a standing wave on a string.
   a. Draw the standing wave if the tension is quadrupled while the frequency is held constant.

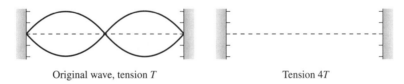

Original wave, tension *T*   Tension 4*T*

   b. Suppose the tension is merely doubled while the frequency remains constant. Will there be
      a standing wave? If so, how many antinodes will it have? If not, why not?

5. This standing wave has a period of 8 ms. Draw snapshot graphs of the string every 1 ms from $t = 1$ ms to $t = 8$ ms. Think carefully about the proper amplitude at each instant.

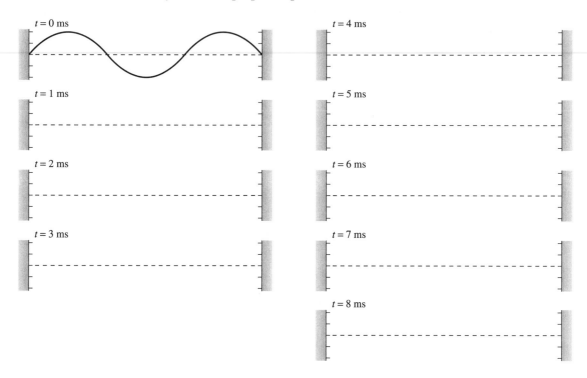

6. The figure shows a standing wave on a string. It has frequency f.

   a. Draw the standing wave if the frequency is changed to $\frac{2}{3}f$ and to $\frac{3}{2}f$.

   Original wave, frequency f Frequency $\frac{2}{3}f$ Frequency $\frac{3}{2}f$

   b. Is there a standing wave if the frequency is changed to $\frac{1}{4}f$? If so, how many antinodes does it have? If not, why not?

21.4 Standing Sound Waves and Musical Acoustics

7. The picture shows a displacement standing sound wave in a 32-mm-long tube of air that is open at both ends.

   a. Which mode (value of m) standing wave is this? _____

   b. Are the air molecules vibrating vertically or horizontally? Explain.

   c. At what distances from the left end of the tube do the molecules oscillate with maximum amplitude?

8. The purpose of this exercise is to visualize the motion of the air molecules for the standing wave of Exercise 7. On the next page are nine graphs, every one-eighth of a period from $t = 0$ to $t = T$. Each graph represents the displacements at that instant of time of the molecules in a 32-mm-long tube. Positive values are displacements to the right, negative values are displacements to the left.

   a. Consider nine air molecules that, in equilibrium, are 4 mm apart and lie along the axis of the tube. The top picture on the right shows these molecules in their equilibrium positions. The dotted lines down the page—spaced 4 mm apart—are reference lines showing the equilibrium positions. Read each graph carefully, then draw nine dots to show the positions of the nine air molecules at each instant of time. The first one, for $t = 0$, has already been done to illustrate the procedure.

   Note: It's a good approximation to assume that the left dot moves in the pattern 4, 3, 0, -3, -4, -3, 0, 3, 4 mm; the second dot in the pattern 3, 2, 0, -2, -3, -2, 0, 2, 3 mm; and so on.

   b. At what times does the air reach maximum compression, and where does it occur?

   Max compression at time _____   Max compression at position _____

   _____   _____

   c. What is the relationship between the positions of maximum compression and the nodes of the standing wave?

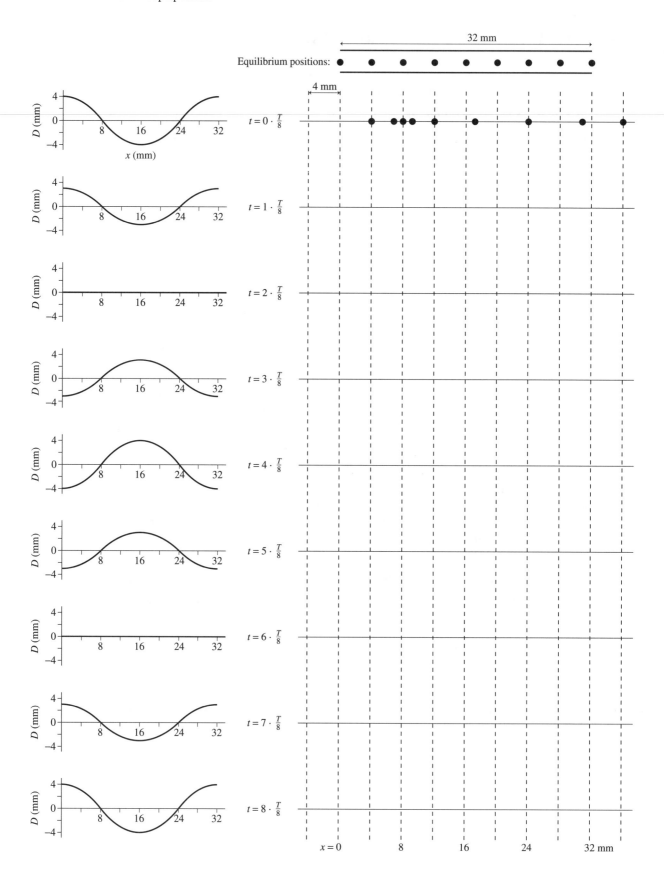

21.5 Interference in One Dimension

21.6 The Mathematics of Interference

9. The figure shows a snapshot graph at $t = 0$ s of loudspeakers emitting triangular-shaped sound waves. Speaker 2 can be moved forward or backward along the axis. Both speakers vibrate in phase at the same frequency. The second speaker is drawn below the first, so that the figure is clear, but you want to think of the two waves as overlapped as they travel along the x-axis.

a. On the left set of axes, draw the $t = 0$ s snapshot graph of the second wave if speaker 2 is placed at each of the positions shown. The first graph, with $x_{\text{speaker}} = 2$ m, is already drawn.

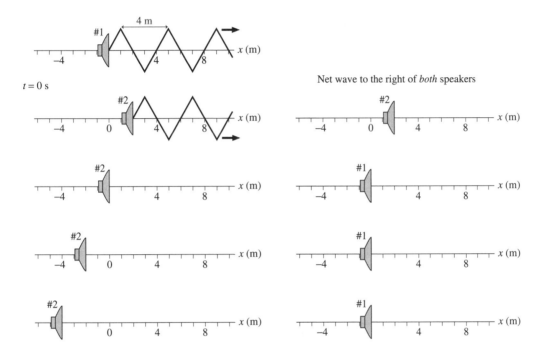

b. On the right set of axes, draw the superposition $D_{\text{net}} = D_1 + D_2$ of the waves from the two speakers. D_{net} exists only to the right of *both* speakers. It is the net wave traveling to the right.

c. What separations between the speakers give constructive interference? _____

d. What are the $\Delta x/\lambda$ ratios at the points of constructive interference? _____

e. What separations between the speakers give destructive interference? _____

f. What are the $\Delta x/\lambda$ ratios at the points of destructive interference? _____

10. Consider the two loudspeakers of Exercise 9.

   a. Copy the speaker 1 and 2 graphs from Exercise 9 onto the first set of axes below for the situation in which speaker 2 is 4 m behind speaker 1. Then draw their superposition on the axes at the right. This simply repeats your last set of graphs from Exercise 9.

   b. On the axes on the left, draw snapshot graphs of the two waves at times $t = \frac{1}{4}T$, $\frac{2}{4}T$, and $\frac{3}{4}T$, where T is the wave's period.

   c. On the right axes, draw the superposition of the two waves.

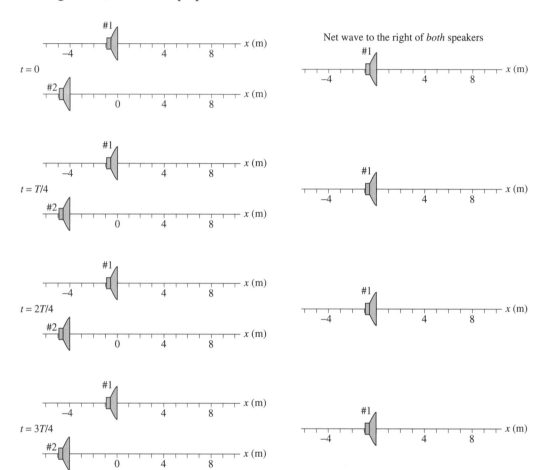

   d. Is the net wave a traveling wave or a standing wave? Use your *observations* to explain.

11. Two loudspeakers are shown at $t = 0$ s. Speaker 2 is 4 m behind speaker 1.

    a. What is the wavelength λ? _____

    b. Is the interference constructive or destructive?

    _____

    c. What is the phase constant ϕ_{10} for wave 1? _____

    What is the phase constant ϕ_{20} for wave 2? _____

    d. At points A, B, C, and D on the x-axis, what are:
    • The distances x_1 and x_2 to the two speakers?
    • The path length difference $\Delta x = x_2 - x_1$?
    • The phases ϕ_1 and ϕ_2 of the two waves at the point (not the phase constant)?
    • The phase difference $\Delta\phi = \phi_2 - \phi_1$?
    Point A is already filled in to illustrate.

	x_1	x_2	Δx	ϕ_1	ϕ_2	$\Delta\phi$
Point A	1 m	5 m	4 m	0.5π rad	2.5π rad	2π rad
Point B						
Point C						
Point D						

e. Now speaker 2 is placed only 2 m behind speaker 1. Is the interference constructive or destructive?

    _____

f. Repeat step c for the points A, B, C, and D.

    _____

	x_1	x_2	Δx	ϕ_1	ϕ_2	$\Delta\phi$
Point A	1 m	3 m	2 m	0.5π rad	1.5π rad	π rad
Point B						
Point C						
Point D						

g. When the interference is constructive, what is $\Delta x/\lambda$? _____ What is $\Delta\phi$? _____

h. When the interference is destructive, what is $\Delta x/\lambda$? _____ What is $\Delta\phi$? _____

12. Two speakers are placed side-by-side at $x = 0$ m. The waves are shown at $t = 0$ s.

    a. Is the interference constructive or destructive?

    b. What is the phase constant ϕ_{10} for wave 1? _____

       What is the phase constant ϕ_{10} for wave 2? _____

    c. At points A, B, C, and D on the x-axis, what are:

       • The distances x_1 and x_2 to the two speakers?
       • The path length difference $\Delta x = x_2 - x_1$?
       • The phases ϕ_1 and ϕ_2 of the two waves at the point (not the phase constant)?
       • The phase difference $\Delta\phi = \phi_2 - \phi_1$?

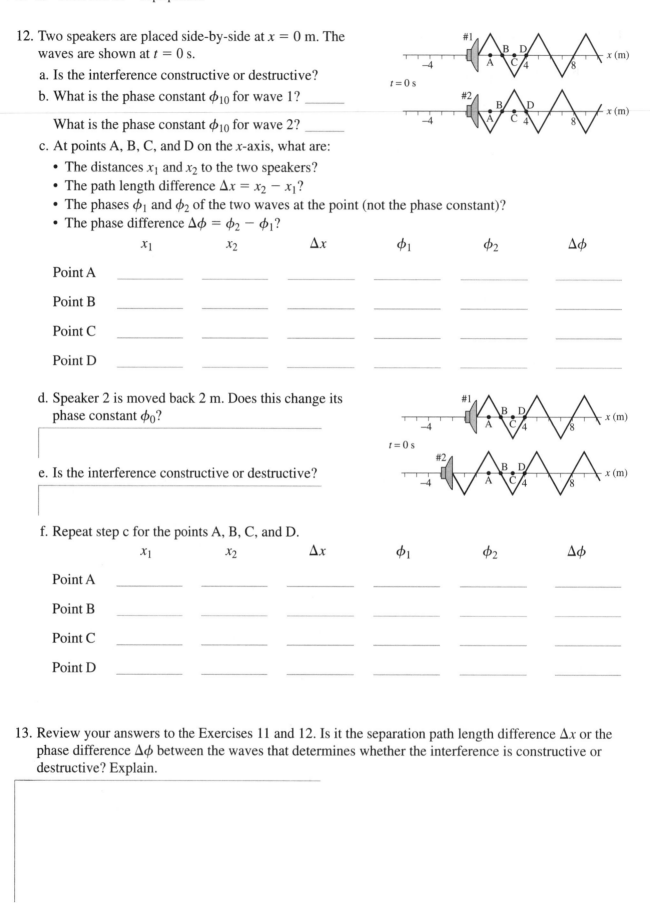

	x_1	x_2	Δx	ϕ_1	ϕ_2	$\Delta\phi$
Point A						
Point B						
Point C						
Point D						

    d. Speaker 2 is moved back 2 m. Does this change its phase constant ϕ_0?

    e. Is the interference constructive or destructive?

    f. Repeat step c for the points A, B, C, and D.

	x_1	x_2	Δx	ϕ_1	ϕ_2	$\Delta\phi$
Point A						
Point B						
Point C						
Point D						

13. Review your answers to the Exercises 11 and 12. Is it the separation path length difference Δx or the phase difference $\Delta\phi$ between the waves that determines whether the interference is constructive or destructive? Explain.

21.7 Interference in Two and Three Dimensions

14. This is a snapshot graph of two plane waves passing through a region of space. Each has a 2 mm amplitude. At each lettered point, what are the displacements of each wave and the net displacement?

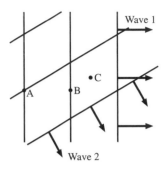

a. Point A: $D_1 =$ _____ $D_2 =$ _____ $D_{net} =$ _____

b. Point B: $D_1 =$ _____ $D_2 =$ _____ $D_{net} =$ _____

c. Point C: $D_1 =$ _____ $D_2 =$ _____ $D_{net} =$ _____

15. Speakers 1 and 2 are 12 m apart. Both emit identical triangular sound waves with $\lambda = 4$ m and $\phi_0 = \pi/2$ rad. Point A is $r_1 = 16$ m from speaker 1.

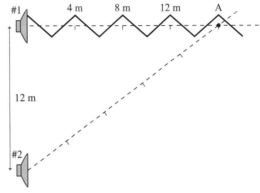

a. What is distance r_2 from speaker 2 to A?

b. Draw the wave from speaker 2 along the dashed line to just past point A.

c. At A, is wave 1 a crest, trough, or zero? _____

At A, is wave 2 a crest, trough, or zero? _____

d. What is the path length difference $\Delta r = r_2 - r_1$? _____ What is the ratio $\Delta r/\lambda$? _____

e. Is the interference at point A constructive, destructive, or in between? _____

16. Speakers 1 and 2 are 18 m apart. Both emit identical triangular sound waves with $\lambda = 4$ m and $\phi_0 = \pi/2$ rad. Point B is $r_1 = 24$ m from speaker 1.

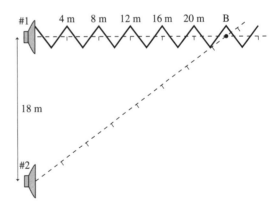

a. What is distance r_2 from speaker 2 to B?

b. Draw the wave from speaker 2 along the dashed line to just past point A.

c. At B, is wave 1 a crest, trough, or zero? _____

At B, is wave 2 a crest, trough, or zero? _____

d. What is the path length difference $\Delta r = r_2 - r_1$? _____ What is the ratio $\Delta r/\lambda$? _____

e. Is the interference at point B constructive, destructive, or in between? _____

17. Two speakers 12 m apart emit identical triangular sound waves with $\lambda = 4$ m and $\phi_0 = 0$ rad. The distances r_1 to points A, B, C, D, and E are shown in the table below.

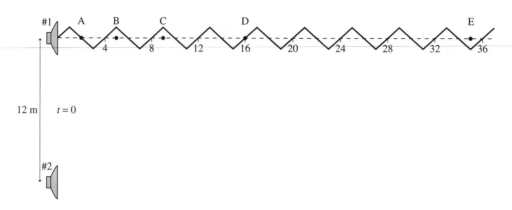

a. For each point, fill in the table and determine whether the interference is constructive (C) or destructive (D).

Point	r_1	r_2	Δr	$\Delta r/\lambda$	C or D
A	2.2 m				
B	5.0 m				
C	9.0 m				
D	16 m				
E	35 m				

b. Are there any points to the right of E, on the line straight out from speaker 1, for which the interference is either exactly constructive or exactly destructive? If so, where? If not, why not?

c. Suppose you start at speaker 1 and walk straight away from the speaker for 50 m. Describe what you will hear as you walk.

18. The figure shows the wave-front pattern emitted by two loudspeakers.

    a. Draw a dot • at points where there is constructive interference. These will be points where two crests overlap *or* two troughs overlap.

    b. Draw an open circle ○ at points where there is destructive interference. These will be points where a crest overlaps a trough.

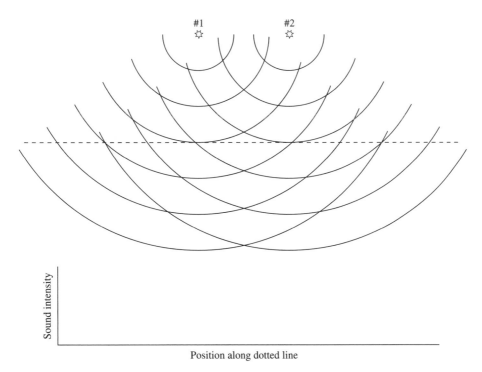

    c. Use a **black** line to draw each "ray" of constructive interference. Use a **red** line to draw each "ray" of destructive interference.

    d. Draw a graph on the axes above of the sound intensity you would hear if you walked along the horizontal dashed line. Use the same horizontal scale as the figure so that your graph lines up with the figure above it.

    e. Suppose the phase constant of speaker 2 is increased by π rad. Describe what will happen to the interference pattern.

21.8 Beats

19. The two waves arrive simultaneously at a point in space from two different sources.

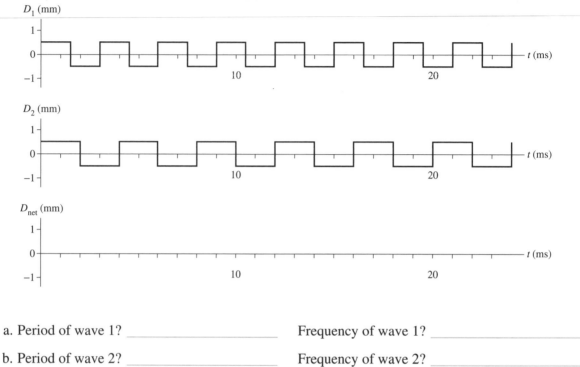

a. Period of wave 1? _____ Frequency of wave 1? _____

b. Period of wave 2? _____ Frequency of wave 2? _____

c. Draw the graph of the net wave at this point on the third set of axes. Be accurate, use a ruler!

d. Period of the net wave? _____ Frequency of the net wave? _____

e. Is the frequency of the superposition what you would expect as a beat frequency? Explain.

22 Wave Optics

22.1 Light and Optics

22.2 The Interference of Light

1. The figure shows the light intensity recorded by a piece of film in an interference experiment. Notice that the light intensity comes "full on" at the edges of each maximum, so this is *not* the intensity that would be recorded in Young's double-slit experiment.

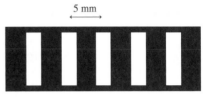

   a. Draw a graph of light intensity versus position on the film. Your graph should have the same horizontal scale as the "photograph" above it.

   b. Is it possible to tell, from the information given, what the wavelength of the light is? If so, what is it? If not, why not?

2. The graph shows the light intensity on the viewing screen during a double-slit interference experiment.

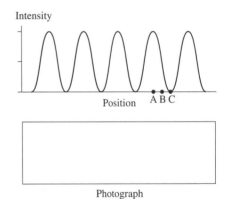

   a. Draw the "photograph" that would be recorded if a piece of film were placed at the position of the screen. Your "photograph" should have the same horizontal scale as the graph above it. Be as accurate as you can. Let the white of the paper be the brightest intensity and a very heavy pencil shading be the darkest.

   b. Three positions on the screen are marked as A, B, and C. Draw history graphs showing the displacement of the light wave at each of these three positions as a function of time. Show three cycles, and use the same vertical scale on all three.

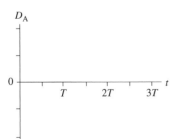

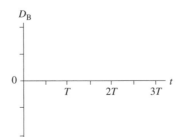

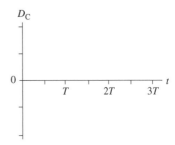

3. The figure below is a double-slit experiment seen looking down on the experiment from above. Although we usually see the light intensity only on a view screen, we can use smoke or dust to make the light visible as it propagates between the slits and the screen. Assuming that the space in the figure is filled with smoke, what kind of light and dark pattern would you see as you look down? Draw the pattern on the figure by shading areas that would appear dark and leaving the white of the paper for areas that would appear bright.

Two slits

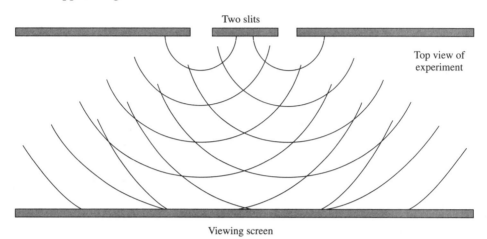

Top view of experiment

Viewing screen

4. The figure shows the viewing screen in a double-slit experiment. For questions a–c, will the fringe spacing increase, decrease, or stay the same? Give an explanation for each.

a. The distance to the screen is increased.

b. The spacing between the slits is increased.

c. The wavelength of the light is increased.

d. Suppose the wavelength of the light is 500 nm. How much farther is it from the dot in the center of fringe E to the more distant slit than it is from the dot to the nearer slit?

22.3 The Diffraction Grating

5. The figure shows four slits in a diffraction grating. A set of Huygens wavelets is spreading out from each slit. Four wave paths, numbered 1 to 4, are shown leaving the slits at angle θ_1. The dashed lines are drawn perpendicular to the paths of the waves.

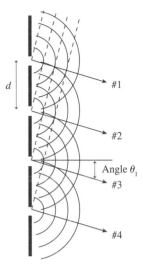

   a. Use a colored pencil or heavy shading to show *on the figure* the extra distance traveled by wave 1 that is not traveled by wave 2.

   b. How many extra wavelengths does wave 1 travel compared to wave 2? Explain how you can tell from the figure.

   c. How many extra wavelengths does wave 2 travel compared to wave 3?

   d. As these four waves combine at some large distance from the grating, will they interfere constructively, destructively, or in between? Explain.

   e. Suppose the wavelength of the light is doubled. (Imagine erasing every other wave front in the picture.) Would the interference at angle θ_1 then be constructive, destructive, or in between? Explain. Your explanation should be based on the figure, not on some equation.

   f. Suppose the slit spacing is doubled. (Imagine closing every other slit in the picture). Would the interference at angle θ_1 then be constructive, destructive, or in between? Again, base your explanation on the figure.

6. These are the same slits as in Exercise 5? Waves with the same wavelength are spreading out on the right side.

   a. Draw four paths, starting at the slits, at an angle θ_2 such that the wave along each path travels *two* wavelengths farther than the next wave. Also draw dashed lines at right angles to the travel direction. Your picture should look much like the figure of Exercise 5, but with the waves traveling at a different angle. Use a ruler!

   b. Do the same for four paths at angle $\theta_{1/2}$ such that each wave travels *one-half* wavelength farther than the next wave.

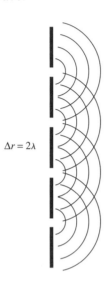

$\Delta r = 2\lambda$

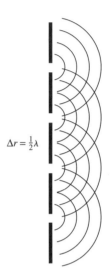

$\Delta r = \frac{1}{2}\lambda$

7. This is the interference pattern on a viewing screen behind two slits. How would the pattern change if the two slits were replaced by 20 slits having the *same spacing d* between adjacent slits?

   a. Would the number of fringes on the screen increase, decrease, or stay the same?

   b. Would the fringe spacing increase, decrease, or stay the same?

   c. Would the width of each fringe increase, decrease, or stay the same?

   d. Would the brightness of each fringe increase, decrease, or stay the same?

22.4 Single-Slit Diffraction

8. Plane waves of light are incident on two narrow, closely-spaced slits. The graph shows the light intensity seen on a screen behind the slits.

   a. Draw a graph on the axes below to show the light intensity on the screen if the right slit is blocked, allowing light to go only through the left slit.

   b. Explain why the graph will look this way.

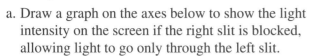

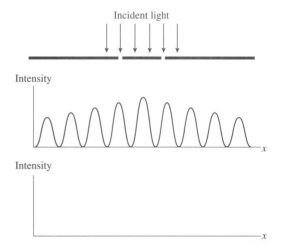

Intensity

Intensity

9. This is the light intensity on a viewing screen behind a slit of width a. The light's wavelength is λ. Is $\lambda < a$, $\lambda = a$, $\lambda > a$, or is it not possible to tell? Explain.

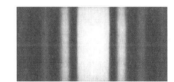

10. This is the light intensity on a viewing screen behind a rectangular opening in a screen. Is the shape of the opening

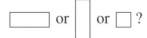 or ⬜ or ▯ ?

Explain.

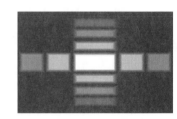

11. The graph shows the light intensity on a screen behind a 0.2-mm-wide slit illuminated by light with a 500 nm wavelength.

    a. Draw a *picture* in the box of how a photograph taken at this location would look. Use the same horizontal scale, so that your picture aligns with the graph above. Let the white of the paper represent the brightest intensity and the darkest you can draw with a pencil or pen be the least intensity.

    b. Using the same horizontal scale as in part a, draw graphs showing the light intensity if

       i. $\lambda = 250$ nm, $a = 0.2$ mm.

       ii. $\lambda = 1000$ nm, $a = 0.2$ mm.

       iii. $\lambda = 500$ nm, $a = 0.1$ mm.

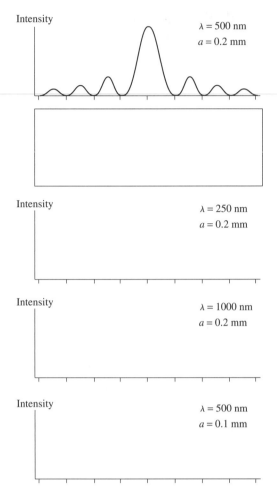

22.5 Circular-Aperture Diffraction

12. This is the light intensity on a viewing screen behind a circular aperture.

    a. If the wavelength of the light is increased, will the width of the central maximum increase, decrease, or stay the same? Explain.

    b. If the diameter of the aperture is increased, will the width of the central maximum increase, decrease, or stay the same? Explain.

    c. How will the screen appear if the aperture diameter is decreased to less than the wavelength of the light?

22.6 Interferometers

13. The figure shows a tube through which sound waves with $\lambda = 4$ cm travel from left to right. Each wave divides at the first junction and recombines at the second. The dots and triangles show the positions of the wave crests at $t = 0$ s—rather like a very simple wave front diagram.

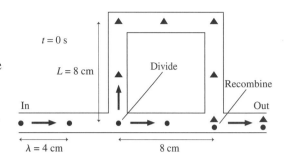

a. Do the recombined waves interfere constructively or destructively? Explain.

b. How much *extra* distance does the upper wave travel? ....

How many wavelengths is this extra distance? ....

c. Below are tubes with $L = 9$ cm and $L = 10$ cm. Use dots to show the wave crest positions at $t = 0$ s for the wave taking the lower path. Use triangles to show the wave crests at $t = 0$ s for the wave taking the upper path. The wavelength is $\lambda = 4$ cm. Assume that the first crest is at the left edge of the tube, as in the figure above.

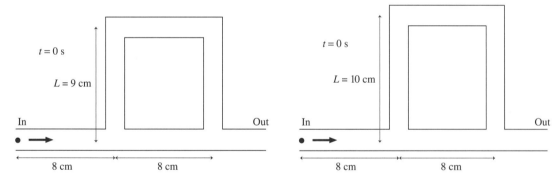

d. How many *extra* wavelengths does the upper wave travel in the $L = 9$ cm tube? ....

What kind of interference does the $L = 9$ cm tube produce? ....

e. How many *extra* wavelengths does the upper wave travel in the $L = 10$ cm tube? ....

What kind of interference does the $L = 10$ cm tube produce? ....

14. A Michelson interferometer has been adjusted to produce a bright spot at the center of the interference pattern.
    a. Suppose the wavelength of the light is halved. Is the center of the pattern now bright or dark, or is it not possible to say? Explain.

    b. Suppose the wavelength of the light doubled to twice its original value. Is the center of the pattern now bright or dark, or is it not possible to say? Explain.

23 Ray Optics

Note: Please use a ruler or straight edge for drawing light rays.

23.1 The Ray Model of Light

1. If you turn on your car headlights during the day, the road ahead of you doesn't appear to get brighter. Why not?

2. a. Draw four or five rays from the object that allow A to see the object.
   b. Draw four or five rays from the object that allow B to see the object.

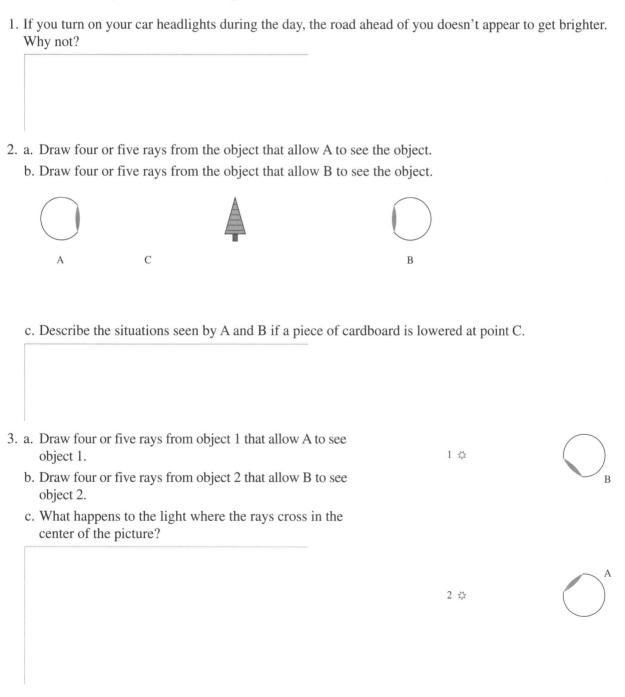

   c. Describe the situations seen by A and B if a piece of cardboard is lowered at point C.

3. a. Draw four or five rays from object 1 that allow A to see object 1.
   b. Draw four or five rays from object 2 that allow B to see object 2.
   c. What happens to the light where the rays cross in the center of the picture?

4. A point source of light illuminates a slit in an opaque barrier.

   a. On the screen, sketch the pattern of light that you expect to see. Let the white of the paper represent light areas; shade dark areas. Mark any relevant dimensions.

   b. What will happen to the pattern of light on the screen if the slit width is reduced to 0.5 cm?

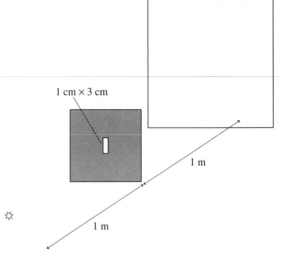

5. In each situation below, light passes through a 1-cm-diameter hole and is viewed on a screen. For each, sketch the pattern of light that you expect to see on the screen. Let the white of the paper represent light areas; shade dark areas.

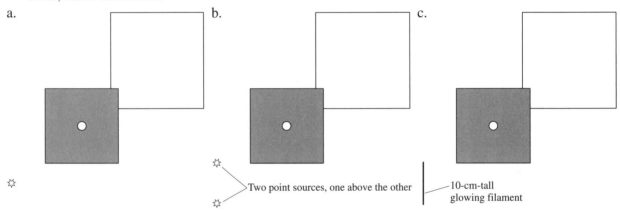

a.

b.

c.

Two point sources, one above the other

10-cm-tall glowing filament

6. Light from a bulb passes through a pinhole. On the screen, sketch the pattern of light that you expect to see.

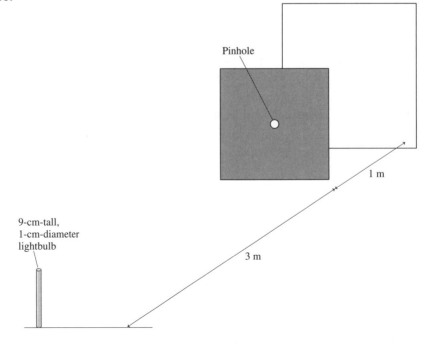

23.2 Reflection

7. a. Draw five rays from the object that pass through points A to E after reflecting from the mirror. Make use of the grid to do this accurately.

   b. Extend the reflected rays behind the mirror.

   c. Show and label the image point.

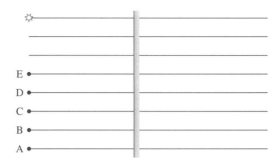

8. a. Draw *one* ray from the object that enters the eye after reflecting from the mirror.

   b. Is one ray sufficient to tell your eye/brain where the image is located?

   c. Use a different color pen or pencil to draw two more rays that enter the eye after reflecting. Then use the three rays to locate (and label) the image point.

   d. Do any of the rays that enter the eye actually pass through the image point?

9. You are looking at the image of a pencil in a mirror.

   a. What happens to the image you see if the top half of the mirror, down to the midpoint, is covered with a piece of cardboard? Explain.

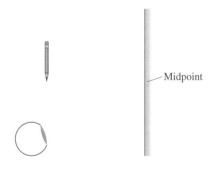

Midpoint

   b. What happens to the image you see if the bottom half of the mirror is covered with a piece of cardboard? Explain.

10. The two mirrors are perpendicular to each other.

a. *Use a ruler* to draw a ray directly from the object to point A. Then draw two rays that strike the mirror about 3 mm (1/8 in) on either side of A. Draw the reflections of these three rays, making sure each obeys the law of reflections, then extend the reflections either forward or backward to locate an image point.

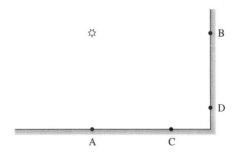

b. Do the same for points B, C, and D.

c. How many images are there, and where are they located?

23.3 Refraction

11. Draw seven rays from the object that refract after passing through the seven dots on the boundary.

a. 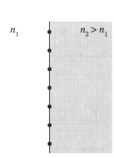 b. 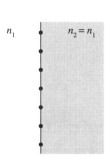 c.

12. Complete the trajectories of these three rays through material 2 and back into material 1. Assume $n_2 < n_1$.

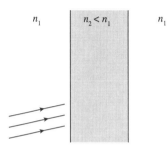

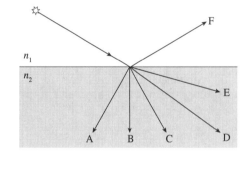

13. The figure shows six conceivable trajectories of light rays leaving an object. Which, if any, of these trajectories are impossible? For each that is possible, what are the requirements of the index of refraction n_2?

Impossible ....

Requires $n_2 > n_1$ ....

Requires $n_2 = n_1$ ....

Requires $n_2 < n_1$ ....

Possible for any n_2 ....

14. Complete the ray trajectories through the two prisms shown below.

a. 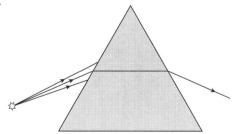 b.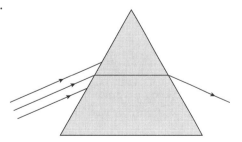

15. Draw the trajectories of seven rays that leave the object heading toward the seven dots on the boundary. Assume $n_2 < n_1$ and $\theta_c = 45°$.

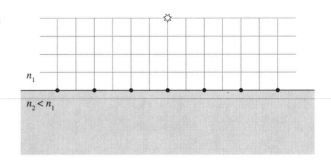

23.4 Image Formation by Refraction

16. a. Draw rays that refract after passing through points B, C, and D. Assume $n_2 > n_1$.

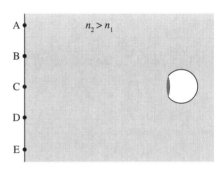

   b. Use dashed lines to extend these rays backward into medium 1. Locate and label the image point.

   c. Now draw the rays that refract at A and E.

   d. Use a different color pen or pencil to draw three rays from the object that enter the eye.

   e. Does the distance to the object *appear* to be larger than, smaller than, or the same as the true distance? Explain.

17. A thermometer is partially submerged in an aquarium. The underwater part of the thermometer is not shown.

   a. As you look at the thermometer, does the underwater part appear to be closer than, farther than, or the same distance as the top of the thermometer?

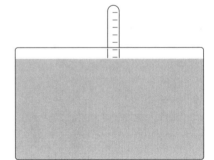

   b. Complete the drawing by drawing the bottom of the thermometer as you think it would look.

23.5 Color and Dispersion

18. A beam of white light from a flashlight passes through a red piece of plastic.

    a. What is the color of the light that emerges from the plastic?

    b. Is the emerging light as intense as, more intense than, or less intense than the white light?

    c. The light then passes through a blue piece of plastic. Describe the color and intensity of the light that emerges.

19. Suppose you looked at the sky on a clear day through pieces of red and blue plastic oriented as shown. Describe the color and brightness of the light coming through sections 1, 2, and 3.

    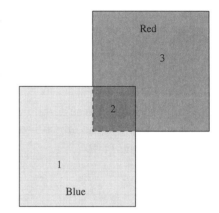

    Section 1:

    Section 2:

    Section 3:

20. Sketch a plausible absorption spectrum for a patch of bright red paint.

    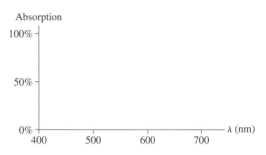

21. Sketch a plausible absorption spectrum for a piece of green plastic.

    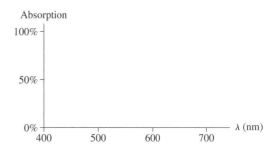

23.6 Thin Lenses: Ray Tracing

22. a. Continue these rays through the lens and out the right side.

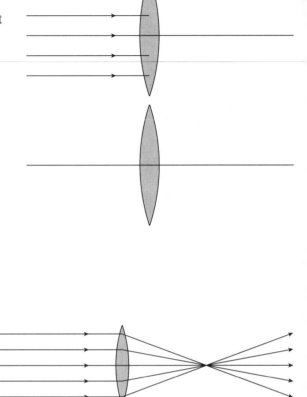

   b. Is the point where the rays converge the same as the focal point of the lens? Or different? Explain.

   c. Place a point source of light at the place where the rays converged in part b. Draw several rays heading left, toward the lens. Continue the rays through the lens and out the left side.

   d. Do these rays converge? If so, where?

23. The top two figures show test data for a lens. The third figure shows a point source near this lens and four rays heading toward the lens.

   a. For which of these rays do you know, from the test data, its direction after passing through the lens?

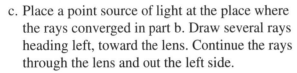

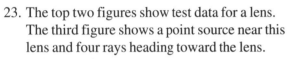

   b. Draw the rays you identified in part a as they pass through the lens and out the other side.

   c. Use a different color pen or pencil to draw the trajectories of the other rays.

   d. Label the image point. What kind of image is this?

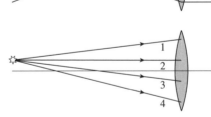

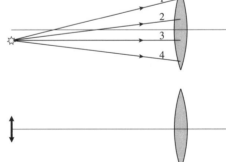

   e. The fourth figure shows a second point source. Use ray tracing to locate its image point.

   f. The fifth figure shows an extended object. Have you learned enough to locate its image? If so, draw it. _____

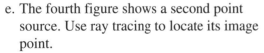

24. An object is near a lens whose focal points are marked with dots.

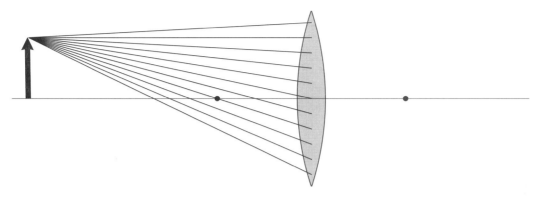

   a. Identify the three special rays and continue them through the lens.

   b. Use a different color pen or pencil to draw the trajectories of the other rays.

25. a. Consider *one* point on an object near a lens. What is the minimum number of rays needed to locate its image point?

   b. For each point on the object, how many rays from this point actually strike the lens and refract to the image point?

26. An object and lens are positioned to form a well-focused, inverted image on a viewing screen. Then a piece of cardboard is lowered just in front of the lens to cover the *top half* of the lens. Describe what happens to the image on the screen. What will you see when the cardboard is in place?

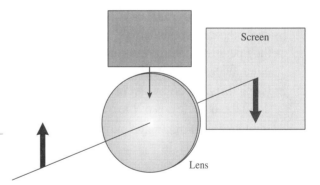

Screen

Lens

27. An object is near a lens whose focal points are shown.

    a. Use ray tracing to locate the image of this object.

    b. Is the image upright or inverted?

    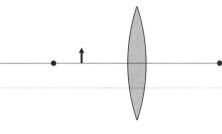

    c. Is the image height larger or smaller than the object height?

    d. Is this a real or a virtual image? Explain how you can tell.

28. The top two figures show test data for a lens. The third figure shows a point source near this lens and four rays heading toward the lens.

    a. For which of these rays do you know, from the test data, its direction after passing through the lens?

    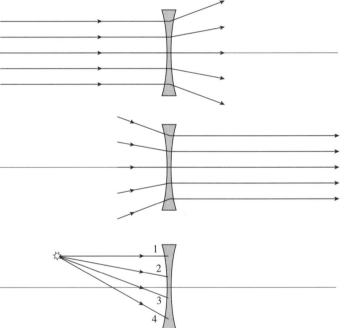

    b. Draw the rays you identified in part a as they pass through the lens and out the other side.

    c. Use a different color pen or pencil to draw the trajectories of the other rays.

    d. Find and label the image point. What kind of image is this?

23.7 Thin Lenses: Refraction Theory

29. Materials 1 and 2 are separated by a spherical surface. For each part:
    i.   Draw the normal to the surface at the seven dots on the boundary.
    ii. Draw the trajectories of seven rays from the object that pass through the seven dots.
    iii. Trace the refracted rays either forward to a point where they converge or backward to a point from which they appear to diverge.

a.   b.

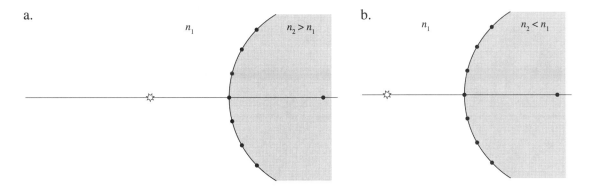

30. A converging lens forms a real image. Suppose the object is moved farther from the lens. Does the image move toward or away from the lens? Explain.

31. A converging lens forms a virtual image. Suppose the object is moved closer to the lens. Does the image move toward or away from the lens? Explain.

23.8 Image Formation with Spherical Mirrors

32. Two spherical mirrors are shown. The center of each is marked. For each:
    i.   Draw the normal to the surface at the seven dots on the boundary.
    ii. Draw the trajectories of seven rays that strike the mirror surface at the dots and then reflect, obeying the law of reflection.
    iii. Trace the reflected rays either forward to a point where they converge or backward to a point from which they diverge.

a.   b.

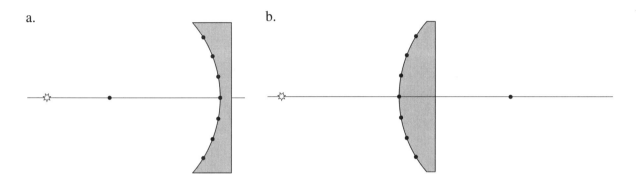

33. An object is placed near a spherical mirror whose focal point is marked.

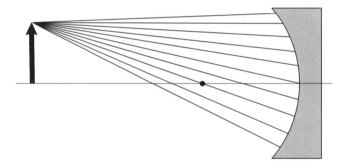

   a. Identify the three special rays and show their reflections.
   b. Use a different color pen or pencil to draw the trajectories of the other rays.

24 Optical Instruments

24.1 Lenses in Combination

24.2 The Camera

1. Use ray tracing to locate the final image of the following two-lens system.

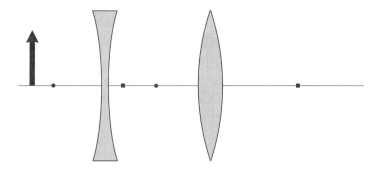

2. Can you tell what's inside the box? Draw one possible combination of lenses inside the box, then show the rays passing through the box.

3. Two converging lenses, whose focal points are marked, are placed in front of an object.

   a. Suppose lens 2 is moved a little to the left. Is the final image of this two-lens system now closer to or farther from lens 2? Explain.

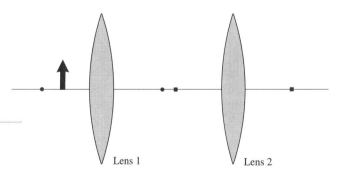

   b. With lens 2 in its original position, suppose lens 1 is moved a little to the left. Is the final image of this two-lens system now closer to or farther from lens 2? Explain.

4. A photographer focuses his camera on an object. Suppose the object moves closer to the camera. To refocus, should the camera lens move closer to or farther from the detector? Explain.

5. The aperture of a camera lens has its diameter halved.

   a. By what factor does the *f*-number change?

   b. By what factor does the focal length change?

   c. By what factor does the exposure time change?

24.3 Vision

6. Two lost students wish to start a fire to keep warm while they wait to be rescued. One student is hyperopic, the other myopic. Which, if either, could use his glasses to focus the sun's rays to an intense bright point of light? Explain.

7. Suppose you wanted special glasses designed to let you see underwater, without a face mask. Should the glasses use a converging or diverging lens? Explain.

24.4 Optical Systems That Magnify

8. a. To double the angular magnification of a magnifier, do you want a lens with twice the focal length or half the focal length? Explain.

   b. Does doubling the angular magnification also double the lateral magnification? Explain.

9. For a telescope, increasing the focal length of the objective increases the overall magnification. For a microscope, increasing the focal length of the objective decreases the overall magnification. Why are they different?

24.5 The Resolution of Optical Instruments

10. A diffraction-limited lens can focus light to a 10-μm-diameter spot on a screen. Do the following actions make the spot diameter larger, smaller, or leave it unchanged?

    a. Decreasing the wavelength of the light: ....

    b. Decreasing the lens diameter: ....

    c. Decreasing the lens focal length: ....

    d. Decreasing the lens-to-screen distance: ....

11. An astronomer is trying to observe two distant stars. The stars are marginally resolved when she looks at them through a filter that passes green light near 550 nm. Which of the following actions would improve the resolution? Assume that the resolution is not limited by the atmosphere.

    a. Changing the filter to a different wavelength? If so, should she use a shorter or a longer wavelength?

    b. Using a telescope with an objective lens of the same diameter but a different focal length? If so, should she select a shorter or a longer focal length?

    c. Using a telescope with an objective lens of the same focal length but a different diameter? If so, should she select a larger or a smaller diameter?

    d. Using an eyepiece with a different magnification? If so, should she select an eyepiece with more or less magnification?

25 Modern Optics and Matter Waves

25.1 Spectroscopy: Unlocking the Structure of Atoms

25.2 X-Ray Diffraction

25.3 Photons

1. The figure shows the spectrum of a gas discharge tube.

What color would the discharge appear to your eye? Explain.

2. The first-order x-ray diffraction of monochromatic x rays from a crystal occurs at angle θ_1. The crystal is then compressed, causing a slight reduction in its volume. Does θ_1 increase, decrease, or stay the same? Explain.

3. Three laser beams have wavelengths $\lambda_1 = 400$ nm, $\lambda_2 = 600$ nm, and $\lambda_3 = 800$ nm. The power of each laser beam is 1 W.

    a. Rank in order, from largest to smallest, the photon energies E_1, E_2, and E_3 in these three laser beams.

    Order:

    Explanation:

   b. Rank in order, from largest to smallest, the number of photons per second N_1, N_2, and N_3 delivered
      by the three laser beams.

   Order:

   Explanation:

4. The top figure is the *negative* of the photograph of a
   single-slit diffraction pattern. That is, the darkest areas
   in the figure were the brightest areas on the screen.
   This photo was made with an extremely large number
   of photons.

   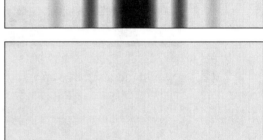

   Suppose the slit is illuminated by an extremely weak
   light source, so weak that only 1 photon passes through
   the slit every second. Data are collected for 60 seconds.
   Draw 60 dots on the empty screen to show how you
   think the screen will look after 60 photons have been
   detected.

5. A light source at point A emits light with a wavelength of 1.0 μm. One photon of light is detected at
   point B, 5.0 μm away from A. On the figure, draw the trajectory that the photon follows from A to B.

   A
   ☆

   B
   •

   5 μm

25.4 Matter Waves

6. The figure is a simulation of the electrons detected behind a very narrow double slit. Each bright dot represents one electron. How will this pattern change if the following experimental conditions are changed? Possible changes you should consider include the number of dots and the spacing, width, and positions of the fringes.

   a. The electron-beam intensity is increased.

   b. The electron speed is reduced.

   c. The electrons are replaced by positrons with the same speed. Positrons are antimatter particles that are identical to electrons except that they have a positive charge.

   d. One slit is closed.

7. Very slow neutrons pass through a single, very narrow slit. Use 50 or 60 dots to show how the neutron intensity will appear on a neutron-detector screen behind the slit.

8. To have the best resolution, should an electron microscope use very fast electrons or very slow electrons? Explain.

25.5 Energy Is Quantized

9. a. For the first few allowed energies of a particle in a box to be large, should the box be very big or very small? Explain.

   b. Which is likely to have larger values for the first few allowed energies: an atom in a molecule, an electron in an atom, or a proton in a nucleus? Explain.

10. The smallest allowed energy of a particle in a box is 2.0×10^{-20} J. What will be the smallest allowed energy if the length of the box is doubled and the particle's mass is halved?